*A Warning from the
Golden Toad*

A Warning from the Golden Toad

TIM FLANNERY

PENGUIN BOOKS — GREEN IDEAS

PENGUIN BOOKS

UK | USA | Canada | Ireland | Australia
India | New Zealand | South Africa

Penguin Books is part of the Penguin Random House group of companies
whose addresses can be found at global.penguinrandomhouse.com.

First published in *The Weather Makers*, Text Publishing, 2005
This extract published in Penguin Books 2021

001

Copyright © Tim Flannery, 2005

Set in 12.5/15pt Dante MT Std
Typeset by Jouve (UK), Milton Keynes
Printed and bound in Great Britain by Clays Ltd, Elcograf S.p.A.

The authorized representative in the EEA is Penguin Random House Ireland,
Morrison Chambers, 32 Nassau Street, Dublin D02 YH68

A CIP catalogue record for this book is available from the British Library

ISBN: 978–0–241–51443–6

www.greenpenguin.co.uk

Contents

The Long Summer

The long summer of the last 8000 years is without doubt *the* crucial event in human history. Although agriculture commenced earlier (around 10,500 years ago in the Fertile Crescent in the Middle East), it was during this period that we acquired most of our major crops and domestic animals, the first cities came into being, the first irrigation ditches were dug, the first words written down, and the first coins minted.

And these changes happened not once, but independently many times in different parts of the world. Before our long summer was 5000 years old, cities had sprung up in Western Asia, East Asia, Africa and central America. The similarities in their temples, homes and fortifications are astonishing.

It is as if the human mind contained a plan for the city all along, and was just waiting until conditions permitted to build it.

These human settlements were ruled by an elite who relied on artisans. In a few societies writing developed, and in even the earliest of these jottings – clay tablets from ancient Mesopotamia – we recognise life as it is lived in a great metropolis.

Did this long summer result from a cosmic fluke? Were Milankovich's cycles, the Sun and Earth all 'just right' to create a warm period of unprecedented stability? During every warm period we know about over the last million years Milankovich's cycles caused a sudden spike in temperature followed by a long, unstable cooling. There is nothing unique about the current Milankovich cycle that can account for the long summer. Indeed, were Milankovich cycles still controlling Earth's climate, we should be feeling a distinct chill by now.

As he tried to explain the long summer, Bill Ruddiman, an environmental scientist at the

University of Virginia, began to look for a unique factor – something that was operating only in this last cycle, but in none of the earlier ones. That unique factor, he decided, was us.

The Nobel laureate Paul Crutzen (awarded the prize for research into the ozone hole) and his colleagues had already recognised and named a new geological Period in honour of our species. They called it the Anthropocene – meaning the age of humanity – and they marked its dawn at 1800 when methane and CO_2, brewed up by the gargantuan machines of the Industrial Revolution, first began to influence Earth's climate.

Ruddiman added a revolutionary twist to this argument: he detected what he believes to be human influences on Earth's climate that occurred long before 1800.

Charting the levels of two critical greenhouse gases – methane and CO_2 – in air bubbles trapped in the Greenland and Antarctic ice sheets, Ruddiman discovered that, beginning 8000 years ago, the Milankovich cycles could not explain what actually happened. Methane should have commenced declining at that time, and gone into a rapid decline by 5000 years ago. Instead, after taking a shallow dip, methane concentrations begin a slow but emphatic rise.

This, Ruddiman argues, is evidence that humans had wrested control of methane emissions from nature, and so we should mark the dawn of the Anthropocene at 8000 years ago rather than 200. It was the beginnings of agriculture – particularly wet agriculture of the kind practised in flooded rice paddies in eastern Asia – that tipped the balance. These agricultural systems can be prodigious producers of methane. Farmers of other crops that require swampy conditions were making their own contributions at around this time. Taro agriculture (which involves the creation and maintenance of watercontrolling structures), for example, was well under way in New Guinea by 8000 years ago.

Even hunter-gatherers may have had a role. The construction of weirs transformed vast areas of southeastern Australia into seasonal swamps. These structures were perhaps the most extensive ever created by non-agricultural people, and were used to regulate swamps for the production of eels. Harvested en masse at great gatherings of the tribes, the eels were then dried and smoked to be traded over large distances.

Ruddiman also found evidence in the ice bubbles that the concentration of CO_2 in the atmosphere was being influenced by humans far earlier than first imagined. CO_2 levels rise rapidly as the glacial stage ends, then typically begin a slow decline towards the next cold period. But in this cycle they kept rising. By 1800 atmospheric CO_2 had risen to 280 parts per million. If natural cycles were still solely in control of Earth's carbon budget, Ruddiman states, CO_2 should have then stood at only 240 parts per million, and be declining.

At first glance his argument looks flimsy. After all, early humans would have needed to emit twice as much carbon as our industrial age did between 1850 and 1990 – an output only made possible by an unprecedented population using coal-burning machines.

The key, notes Ruddiman, is time. Eight thousand years is a long span, and as humans cut and burned forests around the globe their activities acted like a hand casting feathers on a set of scales: eventually enough feathers piled up to tip the balance.

The delicate climatic stability created by humanity over the past 8000 years, Ruddiman argues, was still vulnerable to the great cycles of Milankovich. The archaeologist Brian Fagan argues that these cycles could be amplified into truly monumental impacts on human societies. The slight shift in Earth's orbit between 10,000 and 4000 BC brought between 7 and 8 per cent more sunlight to the Northern Hemisphere.

This changed atmospheric circulation, which resulted in increased rainfall in Mesopotamia by 25 to 30 per cent. What was once a desert was transformed into a verdant plain that supported dense farming communities. After 3800 BC, however, Earth's orbit reverted to its former pattern and rainfall dropped off, forcing many farmers to abandon their fields and wander in search of food.

Fagan believes that the famine-driven wanderers found refuge in a few strategic locations such as Uruk (now in southern Iraq), where irrigation canals branched off the main rivers. Here the starving migrants were put to work by a central authority in construction projects such as the maintenance of canals.

Reduced rainfall, Fagan argues, also forced Uruk's farmers to innovate, and so they used, for the first time, ploughs and animals to till fields in a rotation that involved producing two crops per year.

With grain production localised around strategic towns, surrounding settlements began to specialise in producing goods such as pottery, metals or fish, which were traded at Uruk's markets for the ever-scarcer grain.

Each of these changes led to the development of a more centralised authority, which in turn employed the world's first bureaucrats, whose job it was to tally and distribute the vital grain.

The sum of all of this change was a shift in

human organisation, and by 3100 BC Mesopotamia's southern cities had become the world's first civilisations. Indeed the city, Fagan argues, is a key human adaptation to drier climatic conditions.

Let's return now to Bill Ruddiman's analysis, because it contains several twists in its tail. He sees a clear correlation with times of low atmospheric CO_2 and several plagues caused by the bacterium *Yersinia pestis* – the black plague of medieval times. These epidemics were global in their reach and killed so many people that forests were able to grow back on deserted farmland. In the process they absorbed CO_2, lowering atmospheric concentrations by 5 to 10 parts per million. Global temperatures then fell and periods of relative cold ensued in places such as Europe.

Ruddiman's thesis implies that, by adding sufficient greenhouse gases to delay another ice age, yet not overheating the planet, the ancients performed an act of chemical wizardry. Today, however, the changes scientists are detecting in our atmosphere are so great that time's gates appear once again to be opening.

There are unmistakable signs that the Anthropocene is turning ugly. Will it become the shortest geological Period on record?

Digging up the Dead

We walk on earth,
we look after,
like rainbow sitting on top.
But something underneath,
under the ground.
We don't know.
You don't know.
What do you want to do?
If you touch,
you might get cyclone, heavy rain or flood.
Not just here,
you might kill someone in another place.
Might be kill him in another country.
You cannot touch him.

Big Bill Neidjie, *Gagadju Man*, 2001.

Australia's Aborigines live close to the land, and they have a distinctive way of viewing the world. They tend to see a whole picture. Big Bill Neidjie was a truly wise elder who spent his youth living a tribal life always moving, hunting and gathering. When he tells us about the impact of mining in his Kakadu country he doesn't talk of the mines, the tailings and the poisoned earth. In just a handful of words he describes the great cycle that runs from disturbing the eternal living dreaming of the ancestors to the catastrophe awaiting unborn generations.

The challenge he throws down – 'What do you want to do?' – is discomforting, because my country – Bill's country – is pierced through and through with mines of every type, and more coal is drawn from its innards to be shipped off overseas than from any other place on the planet. Yet Bill has intuited the hidden links between mining, climate change and human wellbeing that scientists struggle to understand in their studies of the greenhouse effect. Bill's challenge remains there to be answered, because we still have a chance to decide our future.

Fossil fuels – oil, coal and gas – are all that remain of organisms that, many millions of years ago, drew carbon from the atmosphere. When we burn wood we release carbon that has been out of atmospheric circulation for a few decades, but when we burn fossil fuels we release carbon that has been out of circulation for eons.

Digging up the dead in this way is a particularly bad thing for the living to do.

In 2002, the burning of fossil fuels released a total of 21 billion tonnes of CO_2 into the atmosphere. Of this, coal contributed 41 per cent, oil 39 per cent, and gas 20 per cent. The energy we liberate when we burn these fuels comes from carbon and hydrogen. Because carbon causes climate change, the more carbon-rich a fuel is, the greater danger it presents to humanity's future. Anthracite, the best black coal, is almost pure carbon. Burn a tonne of it and you create 3.7 tonnes of CO_2.

The fuels derived from oil contain two

hydrogen atoms for every one of carbon in their structure. Hydrogen gives more heat when burned than carbon (and in doing so produces only water), so burning oil releases less CO_2 than coal does.

The fossil fuel with the least carbon content is methane, which has just one carbon atom for every four hydrogen atoms.

These fuels form a stairway leading away from carbon as the fuel for our economy. Even using very advanced methods (and most coal-fired power plants come nowhere near this), burning anthracite to generate electricity results in 67 per cent more CO_2 emissions than does methane, while brown coal (which is younger, and has more moisture and impurities) produces 130 per cent more.

To see the most polluting power plant operating in the world's major industrialised countries (in terms of CO_2), you need to travel to Gippsland, Victoria, where the brown-coal fired Hazelwood power station provides electricity for much of the state.

From a climate change perspective, then, there's a world of difference between burning gas or coal.

Coal is our planet's most abundant fossil fuel. Those in the industry often refer to it as 'buried sunshine', because coal is the fossilised remains of plants that grew in swamps millions of years ago. In places like Borneo you can see the initial stages of the coal-forming process taking place. There, huge trees topple over and sink into the quagmire, where a lack of oxygen impedes rotting. More and more dead vegetation builds up until a thick layer of sodden plant matter is in place. Rivers then wash sand and silt into the swamp, which compresses the vegetation, driving out moisture and other impurities.

As the swamp is buried deeper and deeper in the earth, heat and time alter the chemistry of the wood, leaves and other organic matter to produce decomposing debris called peat. First, the peat is converted to brown coal and, after many millions of years, the brown coal becomes bituminous coal, which has less moisture and impurities. If further pressure and heat are applied, and more impurities removed, it can become anthracite. At its most exquisite anthracite – in the form of jet – is a jewel as beautiful as a diamond.

Certain times in Earth's history have been better for forming coal than others. In the Eocene Period, around 50 million years ago, great swamps lay over parts of Europe and Australia. Their buried remains form the brown coal deposits found today.

Much of the world's anthracite lived during the Carboniferous Period, 360 to 290 million years ago. Named for the immense coal deposits that were then laid down, the Carboniferous world was a very different place from the wetlands of today.

If you'd been able to punt through the ancient swamps of that bygone era, instead of river red gums or swamp cypress, you would have seen gigantic relatives of clubmosses and lycopods (primitive swamp plants), as well as far stranger plants that are now extinct. The scaly, columnar trunks of the *Lepidodendron* grew in dense forests, each trunk two metres in diameter and soaring forty-five metres into the air. They did not branch until near their tips, where a few short, straggly sproutings bore metre-long grassy leaves.

There were no reptiles, mammals or birds in those long ago times. Instead, the stifling, humid growth teemed with insects and their kin. The atmosphere was rich in oxygen. Millipedes grew to two metres in length, spiders reached a metre across, and cockroaches were thirty centimetres long. There were dragonflies whose wingspans approached a metre, while in the waters below lurked crocodile-sized amphibians with huge heads, wide mouths and beady eyes.

In stealing the buried treasure of this alien world we have set ourselves free from the limits of biological production in our own era.

The march towards a fossil-fuel-dependent future started in the England of Edward I. The King himself so detested the smell of coal that in 1306 he banned the burning of it. There are even records of coal burners being tortured, hanged or decapitated. But England's forests were becoming exhausted. In spite of the King, the English became the first Europeans to burn coal on a large scale.

People had no idea what coal was. Many miners believed that it was a living substance that grew underground, and would grow faster if it were smeared with dung. The stench of brimstone (sulphur) that accompanied its burning was a reminder of the torments of hell underground. People also associated coal with the plague.

Despite all this, by 1700 a thousand tonnes per day was being burned in the city of London. An energy crisis soon loomed. England's mines had been dug so deep that they were filling with water. A way of pumping it out had to be found.

The man who discovered how this might be done was a small-town ironmonger, Thomas Newcomen. His device burned coal to produce steam, which was then condensed to create a vacuum that moved a piston which pumped the water. The first Newcomen engine was installed in a Staffordshire coal mine in 1712. Fifty years later, hundreds of them were at work in mines across the nation, and England's coal production had grown to 6 million tonnes per year.

The ingenious James Watt improved on Newcomen's design and, in 1784, Watt's friend William Murdoch produced the first mobile steam engine. From that moment on, the coming century – the nineteenth – was to be the century of coal. No other fuel source could rival it for cooking, heating, industrial purposes and transport. In 1882 when Thomas Edison opened up the world's first electric light power station in lower Manhattan, electricity production was added to coal's portfolio.

More coal is burned today than at any time in the past.

Two hundred and forty-nine coal-fired power plants were projected to be built worldwide between 1999 and 2009, almost half in China. A further 483 will follow in the decade to 2019, and 710 more between 2020 and 2030. About a third of these will be Chinese, and in total they will produce 710 gigawatts of power. A gigawatt is a billion watts of power. The CO_2 they produce will continue to warm the planet for centuries.

If the nineteenth was the century of coal, the twentieth was the century of oil. On 10 January 1901, on a small hill in Texas called Spindletop, Al Hamill was drilling for oil. He had penetrated more than 300 metres into the sandstone below, and by 10.30 in the morning, disgusted by his lack of success, was about to give up. At that moment, 'with a deafening blast, and a great howling roar, thick clouds of methane gas jetted from the hole. Then came the liquid, a column of it, six inches wide. It rocketed hundreds of feet into the winter sky before falling back to earth as a dark rain.' The discovery of oil in such deep strata was new. Oil drove coal from the fields of transport and home heating.

The trouble with oil is that there is far less of it than coal and it's harder to find.

Oil is the product of life in ancient oceans and estuaries. It is composed primarily of the remains of plankton – in particular single-celled plants known as phytoplankton. When the plankton die their remains are carried down to oxygen-free depths, where their organic matter can accumulate without being consumed by bacteria.

The geological process for making oil is as precise as a recipe for making pancakes.

First the sediments containing the phytoplankton must be buried and compressed by other rocks. Then, the absolutely right conditions are needed to squeeze the organic matter out of the source rocks and to transfer it, through cracks and crevices, into a suitable storage layer. This layer must be porous, but above it must lie another rock thick enough to forbid escape.

In addition, the waxes and fats that are the source of oil need to be 'cooked' at between 100–135°C for millions of years. If the temperature exceeds these limits, all that will result is gas, or else the fossil fuel will be lost entirely. The creation of oil reserves is the result of pure chance – the right rocks being cooked in the right way for the correct time.

The house of Saud, the Sultan of Qatar and the other wealthy principalities of the Middle East all owe their good fortune to this geological accident. The conditions in the rocks of their region have been just right to deliver a bonanza of oil. Before it was tapped, just one Saudi Arabian oilfield, the Ghawar, held a seventh of the entire planet's oil reserves.

Until 1961 the world's oil companies were finding more and more oil every year, much of it in the Middle East. Since then the rate of discovery has dwindled, yet rates of use have gone up. By 1995 humans were using an average of 24 billion barrels of oil per year, but an average of only 9.6 billion barrels was discovered. By 2006 oil was over $US70 a barrel. Many analysts are predicting even higher prices, and perhaps shortages, as soon as 2010, which suggests that something else will be needed to power the economies of the twenty-first century.

That 'something else', many in the industry believe, is natural gas, around 90 per cent of which is methane. Thirty years ago gas supplied just 20 per cent of the world's fossil fuel. If current trends continue, by 2025 it will have overtaken oil as the world's most important fuel source. There are proven reserves of gas sufficient to last fifty years. It's looking likely that this will be the century of gas.

In 1900 the world was home to a billion people or so. By 2000 there were 6 billion of us, each using on average four times as much energy as our forefathers did 100 years earlier. In the twentieth century the burning of fossil fuels increased sixteen-fold.

As the researcher Jeffrey Dukes says, all the carbon and hydrogen in fossil fuels was gathered together through the power of sunlight, captured by long-ago plants. He has calculated that approximately 100 tonnes of ancient plant life is required to create four litres of petrol.

This means that, over each year of our industrial age, humans have required several centuries' worth of ancient sunlight to keep the economy going. The figure for 1997 – around 422 years of fossil sunlight – was typical.

More than 400 years of blazing light from the Sun – and we burn it in a single year!

Dukes's ideas have changed the way I look at the world. Now, as I tread the sandstone pavements around Sydney, I feel the power of long-spent sunbeams on my bare feet. Looking at the rock through a magnifying lens I can see the grains whose rounded edges caress my toes, and I realise that each one of the countless billion grains has been shaped by the power of the Sun. Over 300 million years ago that Sun drew water from an ocean which then fell as rain on a distant mountain range. Bit by bit the rock crumbled and was carried into streams, until all that remained were rounded grains of quartz.

A million times more energy must have gone into creating sand grains than has ever gone into all human enterprise. From the soles of my feet to the top of my sun-warmed head, I instantly know what Dukes is saying about fossil sunlight: the past is an abundant land, whose buried riches are fabulous when compared with our daily ration of solar radiation.

The power and seduction of fossil fuels will be hard to leave behind. If humans were to look to biomass (all living things, but in this case particularly plants) as a replacement, we would consume 50 per cent more than we now produce on land. We're already using 20 per cent more than the planet can sustainably provide, so we need to find sustainable, innovative ways to do this.

In 1961 there were just 3 billion people, and they were using half of the total resources that our global ecosystem could sustainably provide. By 1986, our population topped 5 billion, and we were using *all* of Earth's sustainable production.

By 2050, when the population is expected to level out at around 9 billion, we will be using – if they can still be found – nearly two planets' worth of resources. But for all the difficulty we'll experience in finding those resources, it's our waste – particularly the greenhouse gases – that is the limiting factor.

Since the beginning of the Industrial Revolution a global warming of 0.63°C has occurred on our planet. Its principal cause is an increase in atmospheric CO_2 from around three parts per 10,000 to just under four. Most of the increase in the burning of fossil fuels has occurred over the last few decades.

Nine out of the ten warmest years ever recorded have occurred since 1990.

People in Greenhouses Shouldn't Tell Lies

The opposition to reducing emissions of greenhouse gases is most intense in the US. The American energy sector is full of cashed-up businesses that use their influence to combat concern about climate change, to destroy emerging challengers, and to oppose moves towards greater energy efficiency.

In the 1970s the US was a world leader and innovator in energy conservation, photovoltaics (converting light to energy) and wind technology. Today it lags behind other countries in these areas. Over the past two decades some in the fossil-fuel industry have worked tirelessly to prevent the world from taking serious action to combat climate change.

The US coal producers have been centre-stage in this campaign. In the 90s Fred Palmer, now company vice-president at Peabody Energy, the world's largest coal producer, led a campaign that the Earth's atmosphere 'is deficient in carbon dioxide'. Producing more would bring in an age of eternal summer. Rather like the CEO of an arms manufacturer arguing that a nuclear war would be good for the planet, Peabody Energy wanted to create a world with atmospheric CO_2 of around 1000 parts per million.

Palmer's views were the basis for the propaganda video *The Greening of Planet Earth* which promoted the idea of 'fertilising' the world with CO_2 to boost crop yields by 30 to 60 per cent, thus bringing an end to world hunger. While such ludicrous claims could be laughed off by scientists, many people were misled.

On the other hand, some fossil-fuel companies are playing an active role in combating climate change. BP, for instance, has taken a clear-eyed view of climate change and has moved 'beyond petroleum', making a 20 per cent cut in its own CO_2 emissions, and a profit in doing so. BP has now become one of the world's largest producers of photovoltaic cells.

The previous British Prime Minister Tony

Blair had a firm grasp of the science surrounding the issue. He has described global warming as 'a challenge so far-reaching in its impact and irreversible in its destructive power, that it alters radically human existence . . . There is no doubt that the time to act is now.'

By 2003 Britain's CO_2 emissions had fallen to 4 per cent below what they were in 1990. Significant milestones of this period include the establishment of the Carbon Trust (which helps business address energy use), an obligation by power suppliers to provide 15.4 per cent of their energy from renewable sources, and significant investments in developing wave and tidal power. Britain is also considering expanding its nuclear power capacity.

These debates about how to transfer from fossil fuels to renewable sources of energy will only grow more intense.

Can we find solutions to the problem of global warming while continuing to use fossil fuels?

The coal industry is promoting the idea of pumping CO_2 underground in order to take it out of the atmosphere. The process, known as geosequestration – it means hiding in the earth – is simple in its approach: the industry would bury the carbon that it had dug up.

Oil and gas companies have been pumping CO_2 underground for years. A good example is the Norwegian Sleipner oilfield in the North Sea where about a million tonnes of CO_2 is pumped underground each year. The Norwegian government has placed a US$40 per tonne tax on CO_2 emissions. This provides the incentive at Sleipner to separate out the CO_2 that comes up with the oil and pump it back into the rocks.

At a few other wells around the world, the CO_2 is pumped back into the oil reserve, helping to maintain head pressure, which assists with the recovery of oil and gas, making the entire operation more profitable. The companies claim 'most' of the CO_2 stays underground. Applying this model to the coal industry, however, is not straightforward.

The problems for coal commence at the smokestack. The stream of CO_2 emitted there is relatively dilute, making its capture unrealistic. The coal industry is promoting a new process known as coal gasification, which creates a more concentrated stream of CO_2 for capture and burial. These plants are not cheap to run: around one quarter of the energy they produce is consumed just in keeping them operating. Building them on a commercial scale will be expensive and it will take decades for them to make a big contribution to power production.

Let's assume that some plants are built and the CO_2 they emit is captured. For every tonne of anthracite burned, around 3.7 tonnes of CO_2 is generated, all of which must be stored. The rocks that produce coal are not often useful for storing CO_2, so the gas would have to be transported away from the power stations. In the case of Australia's Hunter Valley coal mines, for example, it would need to be carried over Australia's Great Dividing Range and hundreds of kilometres to the west to a suitable site.

Once the CO_2 arrives at its destination it must be compressed into a liquid so it can be injected into the ground – a step that typically consumes 20 per cent of the energy yielded by burning coal in the first place. Then a kilometre-deep hole must be drilled and the CO_2 injected. From that day on, the geological formation must be closely monitored. If the gas were ever to escape, it has the potential to kill.

Miners used to call concentrated CO_2 'choke-damp', an appropriate name as it instantaneously smothers its victims.

The largest recent disaster caused by CO_2 occurred in 1986, in Cameroon, central Africa. A volcanic crater-lake known as Nyos belched bubbles of CO_2 into the still night air and the gas settled around the lake's shore. It killed 1800 people and countless thousands of animals, both wild and domesticated.

No one is suggesting we bury CO_2 in volcanic regions like Nyos, so the CO_2 dumps created by industry are unlikely to cause a similar disaster.

Still, Earth's crust is not a purpose-built vessel for holding CO_2, and the storage must last thousands of years. The risk of a leak must be taken seriously.

The amount of CO_2 we would need to bury is mind-boggling. We can use a country like Australia, with its comparatively small population, as an example. Imagine a pile of 200-litre drums, ten kilometres long and five kilometres across, stacked ten drums high. That would be more than 1.3 billion drums, the number required to hold the CO_2 that pours out of Australia's twenty-four coal power stations, which provide power to 20 million people *every day*. Even when compressed to liquid form, that daily output would take up a third of a cubic kilometre, and Australia accounts for less than 2 per cent of global emissions!

Imagine injecting 20 cubic kilometres of liquid CO_2 into the Earth's crust every day of the year for the next century or two.

If we were to try to bury all the emissions from coal, the world would very quickly run out of A-grade reservoirs near power stations.

There are enough fossil fuel reserves on planet Earth to create 5000 billion tonnes of CO_2. How could Earth tuck that away without suffering fatal indigestion?

The best-case scenario for geosequestration is that it will play a small role (at most perhaps 10 per cent by 2050) in the world's energy future.

There are other forms of sequestration – of hiding carbon – which are vital for the future of the planet, and which carry no risk. Earth's vegetation and soils are reservoirs for huge volumes of carbon, and are critical elements in the carbon cycle. Today the world is mostly deforested and its soils exhausted, but soil carbon can be enhanced by following sustainable agricultural and animal farming practices.

This increases the vegetable mould (mostly carbon) in the soil. Lots of carbon – around 1180 gigatonnes – is currently stored this way; more than twice as much as is stored in living vegetation (493 gigatonnes). There is real hope for progress here, in initiatives that include everything from organic market gardening to sustainable rangelands management.

Can we store carbon in forests and long-lived forest products? This involves either planting forests, or preserving them. The Costa Rican government saved half a million hectares of tropical rainforest from logging. This brought it carbon credits equivalent to the amount of CO_2 that would have entered the atmosphere if the forests had been disturbed.

Another example is BP's initiative to fund the planting of 25,000 hectares of pine trees in Western Australia to offset emissions from its refinery near Perth.

Forestry plantations are destined to be cut and used, but they can be a good short-term store for carbon because the furniture and housing they produce are long-lived, and because the roots of the felled trees (along with their carbon) stay in the ground.

The carbon in coal has been safely locked away for hundreds of millions of years, and will remain there for millions more if we refuse to dig it up.

Carbon locked away in forests or the soil is unlikely to remain out of circulation for more than a few centuries. By trading coal storage for tree storage of carbon, we are exchanging a long-term guarantee for a quick fix.

Scientists continue to work on the problem of safe, secure storage for carbon, and perhaps a solution will eventuate. Meanwhile, the competition from less carbon-dense fuels is looking simpler and cheaper by the day.

2050: The Great Stumpy Reef?

Of all the ocean's ecosystems, none is more diverse or beautiful in colour and form than a coral reef. And none, the climate experts and marine biologists tell us, is more endangered by climate change.

Are the world's coral reefs really on the brink of collapse?

It's a question that matters to humanity, for coral reefs yield around US$30 billion in income each year, mostly to people who have few other resources.

But financial loss may prove to be a small thing. The citizens of five nations live entirely on coral atolls, while fringing reefs are all that stand between the invading sea and tens of millions of other people. Destroy these fringing reefs, and for many Pacific nations you have done the equivalent of bulldozing Holland's dykes.

One of every four inhabitants of the oceans spends at least part of its life cycle in coral reefs. Such biodiversity is made possible by both the complex architecture of the corals, which provide many hiding places, and the lack of nutrients in the clear, tropical water.

Low levels of nutrients can promote great diversity. The best example of this is seen on the infertile sand plains of South Africa's Cape Province, where 8000 species of shrubby flowering plants co-exist in a mix as diverse as that of most rainforests.

The coral reefs are the marine equivalent of South Africa's sand plains. The arch-enemy of coral reefs are nutrients, and disturbances that break down the structure of the reefs. Then only a few weedy species – mostly marine algae – can proliferate.

When Alfred Russel Wallace sailed into

Ambon Harbour in what is now eastern Indonesia in 1857, he saw:

> one of the most astonishing and beautiful sights
> I have ever beheld. The bottom was absolutely
> hidden by a continuous series of corals, sponges,
> actiniae, and other marine productions, of mag-
> nificent dimensions, varied forms, and brilliant
> colours. The depth varied from about twenty
> to fifty feet, and the bottom was very uneven,
> rocks and chasms, and little hills and valleys,
> offering a variety of stations for the growth of
> these animal forests. In and out among them
> moved numbers of blue and red and yellow
> fishes, spotted and banded and striped in the
> most striking manner, while great orange or
> rosy transparent medusae floated along near
> the surface. It was a sight to gaze at for hours,
> and no description can do justice to its surpass-
> ing beauty and interest.

During the 1990s I often sailed down Ambon Harbour, yet saw no coral gardens, no medusae, no fishes, nor even the bottom. Instead, the stinking opaque water was thick with effluent and garbage. As I neared the town it just got worse, until I was greeted with rafts of faeces, plastic bags, and the intestines of butchered goats.

Ambon Harbour is just one among countless examples of coral reefs that have been devastated over the course of the twentieth century. Today, the practice of overfishing – including fishing with explosives and poisons – threatens reef survival. Disturbing reef biodiversity can also lead to outbreaks of plague species, such as the crown of thorns starfish. Another problem is the runoff of nutrients from land-based agriculture and polluted cities, which has helped degrade even protected places such as Australia's Great Barrier Reef.

During the 1997–98 El Niño, when the rainforests of Indonesia burnt like never before, the air was thick for months with a smog cloud rich in iron. Before those fires, the coral reefs of southwestern Sumatra were among the most diverse in the world, boasting more that 100 species of hard corals, including massive individuals over a century old. Then, late in 1997, a 'red tide' appeared off Sumatra's coast. The colour was the result of a bloom of minute organisms that fed on the iron in the smog. The toxins they produced caused so much damage it will take the reefs decades to recover, if indeed they ever do.

The smog cloud generated over southeast Asia during the 2002 El Niño was even larger – it was the size of the United States. On such a scale smog can cut sunlight by 10 per cent, and heat the lower atmosphere and ocean. Algal blooms are now devastating coastlines from Indonesia to South Korea and causing hundreds of millions of dollars worth of damage to aquaculture and corals alike. Recovery seems unlikely for any east Asian coral reef.

High temperatures lead to coral bleaching. To understand how we need to examine a reef far from human interference, where warm water alone is causing change. Myrmidon Reef lies far off the coast of Queensland, and the only people who go there are the scientists who survey it every three years. When they last went, in 2004, it looked 'as though it's been bombed'.

This was the result of the reef crest being severely bleached, leaving a forest of dead, white coral. Only on the deeper slopes did life survive.

Coral bleaching occurs whenever sea temperatures exceed a certain threshold. Where the hot water pools the coral turns a deathly white. If the heating is temporary the coral may slowly recover, but when the heat persists it dies. Coral bleaching was little heard of before 1930, and it remained a small-scale phenomenon until the 1970s. It was the 1998 El Niño that triggered the global dying.

The Great Barrier Reef is the most vulnerable reef in the world to climate change. In all, 42 per cent of it was bleached in 1998, with 18 per cent suffering permanent damage.

In 2002, with the renewal of El Niño conditions, a pool of warm water around half a million square kilometres developed over the Great Barrier Reef. This triggered another massive bleaching event that on some inshore reefs killed 90 per cent of all reef-forming corals, and left 60 per cent of the Great Barrier Reef affected. In the few patches of cool water which remained, the coral was undamaged.

And 2006 looked liked it was going to be another dreadful year for the reef, but then Cyclone Larry arrived. It took enough heat from the ocean to stall the bleaching event, using the heat energy to power devastating winds that damaged or destroyed 50,000 homes in Queensland. It was a terrible price to pay to secure the reef, for at least another year.

A panel of seventeen of the world's leading coral reef researchers warned that by 2030 catastrophic damage will have been done to the world's reefs, and by 2050 even the most protected of reefs will be showing massive signs of damage.

According to reef scientists, a further rise of 1°C in global temperature will cause 82 per cent of the Great Barrier Reef to bleach and die; 2°C will bleach 97 per cent of it; and after 3°C there will be 'total devastation'.

It takes the oceans around three decades to

catch up with the heat accumulated in the atmosphere, so it may be that four-fifths of the Great Barrier Reef is one vast zone of the living dead – just waiting for time and warm water to catch up with it.

Extinctions caused by climate change are almost certainly under way on the world's reefs, and a tiny species of coral reef-dwelling fish known as *Gobiodon* species C may be symbolic of them. Most of the habitat used by this tiny creature was destroyed by bleaching during the 1997–98 El Niño, and it can now be seen only on one patch of coral in one lagoon in Papua New Guinea.

'Species C' indicates that it has not yet been formally named, and it may become extinct before this can happen. It isn't an exaggeration to say that we need to multiply the loss of this one little fish a thousandfold to gain a sense of the cascade of extinctions that is occurring right now.

A survey conducted in 2003 revealed that live coral cover had dropped to less than 10 per cent on half of the area of the Great Barrier

Reef. Significant damage was evident in even the healthiest sections. Public outrage made political action inevitable, and the Australian government announced that 30 per cent of the reef would be protected. This meant that commercial fishing would be banned, and other human activities severely curtailed, in the newly protected zone.

But it is not fishing or tourists that is killing the Great Barrier Reef. It is spiralling CO_2 emissions. And Australians produce more CO_2 per person than the people of any other nation on Earth.

If we are to have a chance of saving these wonders of the natural world we need to reduce our greenhouse gas emissions now.

Peril at the Poles

In the final days of 2004, the cities of the world received some astonishing news: beginning at its northern tip, Antarctica was turning green.

Antarctic hair-grass usually survives as sparse tussocks crouched behind the north face of a boulder or some other sheltered spot. Over the southern summer of 2004, however, great green meadows began to appear in what was once the home of the blizzard. It's an emblem of the transformations occurring at the polar ends of our Earth. Yet changes on land are insignificant compared to those occurring at sea, for the sea ice is disappearing.

The subantarctic seas are some of the richest on Earth, despite an almost total absence of the nutrient iron. The presence of sea ice somehow compensates for this: the semi-frozen edge between salt water and the floating ice promotes remarkable growth of the microscopic plankton that is the base of the food chain.

Despite the months of winter darkness the plankton thrive under the ice, allowing the krill that feed on them to complete their seven-year life cycle. And wherever there is krill in abundance there are likely to be penguins, seals and great whales.

Ever since 1976 the krill have been in sharp decline, reducing at the rate of nearly 40 per cent per decade. As the krill numbers have decreased, those of another major grazing species – the jelly-like salps – have increased. Salps were previously confined to more northerly waters. They don't require a great density of plankton to thrive; they can survive on the meagre pickings in the ice-free parts of the Southern Ocean. But salps are so devoid of nutrients that none of the Antarctic's marine mammals or birds finds it worthwhile feeding on them.

The reduction in krill numbers seems to coincide closely with the warming of the ocean and the reduction of sea ice. There is little doubt that climate change is damaging the world's most productive ocean, as well as the largest creatures that exist and that feed there.

Imagine what it would mean for the beasts of the Serengeti in Africa if their grasslands had been reduced by 40 per cent each decade since 1976? Imagine what it would mean if your own living space were slashed by 40 per cent each decade?

The emperor penguin population is now half what it was thirty years ago, while the number of Adelie penguins has declined by 70 per cent.

Southern right whales have only recently begun to return to Australian and New Zealand shores, but they will no longer come, because they need to fatten up on winter krill if they are to travel to their birthing grounds in warmer waters. The humpbacks that traverse the world's oceans will no longer be able to fill their huge bellies, nor will the seals and penguins that frolic in southern seas.

Instead we'll have a defrosting cryosphere (the term scientists use to describe the icy areas of the Earth) and an ocean full of jelly-like salps.

The Antarctic is a frozen continent surrounded by an immensely rich ocean. The Arctic, on the other hand, is a frozen ocean almost entirely surrounded by land. It's also home to 4 million people. Most of the Arctic's inhabitants live on the fringe, and it's there, in places such as southern Alaska, that winters are 2°C to 3°C warmer than they were thirty years ago.

Among the most visible impacts of climate change anywhere on Earth are those caused by the spruce bark beetle. Over the past fifteen years it has killed some 40 million trees in southern Alaska, more than any other insect in North America's recorded history. Two hard winters are usually enough to control beetle numbers, but a run of mild winters in recent years has allowed them to explode.

Collared lemmings are superbly adapted to

life in the cryosphere, for they survive even on the hostile northern coast of Greenland. They're the only rodents whose coat turns white in winter, and whose claws grow into two-pronged shovels for tunnelling through snow. They are so abundant they can migrate en masse in search of food, though it's not true that they commit suicide by running off cliffs.

Scientists predict that, if global warming trends persist, forests will expand northwards to the edge of the Arctic Sea and destroy the vast plains and frozen subsoil of the tundra. Several hundred million birds migrate to these regions to breed. As the forest moves north the great flocks look set to lose more than 50 per cent of their nesting habitat this century alone.

For the collared lemming, the tundra and life itself are inseparable. Experts say it will be extinct before the year 2100. Perhaps all we will have then is a folk memory of the small, suicidal rodent.

But the tragedy will be that the lemmings didn't jump. They were pushed.

The caribou (or reindeer as the species is known in Eurasia) is vital to the Inuit, the Arctic's indigenous people. The Peary caribou is a small, pale subspecies found only in west Greenland and Canada's Arctic islands. A season in the Arctic with less snow but more rain can be devastating. Autumn rains now ice over the lichens that are the creature's winter food supply, causing many to starve. The number of Peary caribou dropped from 26,000 in 1961 to 1000 in 1997. In 1991 it was classified as endangered, which meant that it couldn't be hunted, and so it became irrelevant to the Inuit economy.

The Saami people of Finland have noted a similar icing of the reindeer's winter food supply. As climate change advances, it seems that the Arctic will no longer be a suitable habitat for caribou.

Can we imagine the North Pole without reindeer?

If anything symbolises the Arctic it is surely *nanuk,* the great white polar bear. He is a wanderer and a hunter, and a fair match for man in the white infinity of his polar world. Every inch of the Arctic lies within his grasp: he has been sighted two kilometres up on the Greenland ice cap, and purposefully striding the ice within 150 kilometres of the true Pole itself. For polar bears, having sufficient food to live means lots of sea ice. And the sea ice is disappearing at the rate of 8 per cent per decade.

Polar bears, it's true, will catch lemmings or scavenge dead birds if the opportunity presents itself, but it's sea ice and *netsik* – the ringed seal that lives and breeds there – that are at the core of the *nanuk* economy.

Netsik is the most abundant mammal of the far north and at least 2.5 million of them swim in its berg-cooled seas. Yet at times climatic conditions prevent them from breeding. In 1974 too little snow fell over the Amundsen Gulf for the seals to construct their snow-covered dens on the sea ice. So they left, some travelling as far as Siberia.

And the polar bears? Those that had enough fat to migrate followed the seals, but many could not keep up and starved.

The harp seals live in the Gulf of St Lawrence. This seal population is genetically separate from the rest of the species. Like the ringed seals, they can raise no pups when there is little or no sea ice present – which happened to them in 1967, 1981, 2000, 2001 and 2002. The run of pupless years that opened this century is worrying. When a run of ice-free years exceeds the reproductive life of a female seal – perhaps a dozen years at most – the Gulf of St Lawrence population will become extinct.

The great white bears are already slowly starving as each winter becomes warmer than the one before. A long-term study of 1200 individuals living around Hudson Bay reveals that they are already 15 per cent skinnier on average than they were a few decades ago.

With each year, starving females give birth to fewer cubs. Some decades ago triplets were common; they are now unheard of. And back then around half the cubs were weaned and feeding themselves at eighteen months, while today it's less than one in twenty. In some areas increasing winter rain may collapse birthing dens, killing both the mother and cubs sleeping within. And the early break-up of the ice can separate denning and feeding areas; if the young cubs cannot swim the distances to find food, they will starve to death. In the spring of 2006, for the first time Inuit began to find drowned polar bears: the ice is now too far from shore.

In creating an Arctic with dwindling sea ice, we are creating a monotony of open water and dry land. Without ice, snow and *nanuk,* what will it mean to be Inuit – the people who named the great white bear, and who understand him like no other? When *nanuk* is fit and well-fed he will strip the blubber from a fat seal, leaving the rest to the arctic fox, the raven, and the gulls.

As the Arctic fills with hungry white bears, what will become of these other creatures? The ivory gull has already declined by 90 per cent in Canada over the past twenty years. At that rate, it will not see out the century. The *nanuk* is already on the way to joining the list of endangered species.

It looks as if the loss of the *nanuk* may mark the beginning of the collapse of the entire Arctic ecosystem.

If nothing is done to limit greenhouse gas emissions, it seems certain that around 2050 a day will dawn when no summer ice will be seen in the Arctic – just a vast, dark, turbulent sea. But before the last ice melts the bears will have lost their den sites, feeding grounds and migration corridors.

Perhaps a group of elderly bears will linger on, each year becoming thinner than the last. Or perhaps a dreadful summer will arrive when the denning seals are nowhere to be found. A few bears might survive for a time on a diet of lemming, carrion and sea-caught seals, but they'll be so thin that they will not wake from winter's sleep. So fast are the changes that there are likely to be few or no polar bears in the wild by around 2030.

The changes we're witnessing at the Poles are of the runaway type. Unless we act quickly the realm of the polar bear, the narwhal and the walrus will be replaced by the cold, ice-free oceans of the north, and by the great temperate forests of the Taiga (the largest habitat on Earth, that stretches across Canada, Europe and Asia).

You might think that the encroaching forests, by taking in CO_2 as they grow, would help slow climate change. Scientists estimate that this will be offset by the loss of *albedo* or whiteness. A dark green forest absorbs far more sunlight, and thus captures far more heat, than does snow-covered tundra. The overall impact of foresting the world's northern regions will be to heat our planet even more swiftly.

Once this has happened, no matter what humanity does about its greenhouse gas emissions, it will be too late for a reversal. After existing for millions of years, the north Polar cryosphere will have vanished forever.

A Warning from the Golden Toad

In our story we are yet to meet a single species that has definitely become extinct because of climate change. In the regions where it is likely to have occurred, such as New Guinea's forests and coral reefs, there's been no biologist on hand to document the event. In contrast, there are many researchers at the Monteverde Cloud Forest Preserve in Costa Rica, Central America, where the Golden Toad Laboratory for Conservation is located.

Soon after our fragile planet passed through the climatic magic gate of 1976, abrupt and strange events were observed by the ecologists who spend their lives working in these pristine forests.

During the winter dry season of 1987, the frogs that live in the mossy rainforests one and a half kilometres above the sea began to disappear. Thirty of the fifty species known to inhabit the 30-square-kilometre study site vanished. Among them was a spectacular toad the colour of spun gold. The golden toad lived only on the upper slopes of the mountain. At certain times of the year crowds of the brilliant males gathered around puddles on the forest floor to mate.

The golden toad was discovered and named in 1966, although the Indians knew about it long before. They have myths about a mysterious golden frog that is very difficult to find, but should anyone search the mountains for long enough to find one they will obtain great happiness. Their stories tell of one man who found the frog but let it go because he found happiness too painful to bear. Another released the creature because he didn't recognise happiness when he had it.

Only the males are golden; the females are mottled black, yellow and scarlet. For much of the year it's a secretive creature, spending its time in burrows amid the mossy roots of the woodland. Then, as the dry season gives way to the wet in April-May, it appears above ground en masse, for just a few days or weeks. With such a short time to reproduce, the males fight with each other for top spot and take every opportunity to mate – even if it's only with a field worker's boot.

In her book *In Search of the Golden Frog*, amphibian expert Marty Crump tells us what it was like to see the creature in its mating frenzy:

> I trudge uphill . . . through cloud forest, then through gnarled elfin forest . . . At the next bend I see one of the most incredible sights I've ever seen. There, congregated around several small pools at the bases of dwarfed, windswept trees, are over one hundred DayGlo golden orange toads poised like statues, dazzling jewels against the dark brown mud.

On 15 April 1987 Crump made a note in her field diary that was to have historic significance:

> We see a large orange blob with legs flailing in all directions: a writhing mass of toad flesh.

Closer examination reveals three males, each struggling to gain access to the female in the middle. Forty-two brilliant orange splotches poised around the pool are unmated males, alert to any movement and ready to pounce. Another fifty-seven unmated males are scattered nearby. In total we find 133 toads in the neighbourhood of this kitchen sink-sized pool.

On 20 April:

Breeding seems to be over. I found the last female four days ago, and gradually the males have returned to their underground retreats. Every day the ground is drier and the pools contain less water. Today's observations are discouraging. Most of the pools have dried completely, leaving behind desiccated eggs already covered in mold. Unfortunately, the dry weather conditions of El Niño are still affecting this part of Costa Rica.

As if they knew the fate of their eggs, the toads attempted to breed again in May. This was, as far as the world knows, the last great toad orgy ever to occur. Despite the fact that 43,500 eggs were deposited in the ten pools Crump studied, only twenty-nine tadpoles survived for longer than a week, because the pools once again quickly dried.

The following year Crump was back at Monteverde for the breeding season, but this time things were different. After a long search, on 21 May she located a single male. By June, and still searching, Crump was worried: 'the forest seems sterile and depressing without the bright orange splashes of colour . . . I don't understand what's happening. Why haven't we found a few hopeful males, checking out the pools in anticipation?'

A year was to pass before, on 15 May 1989, a solitary male was again sighted. As it was sitting just three metres from where Crump made her sighting twelve months earlier, it was almost certainly the same toad.

For the second year running he held a lonely vigil, waiting for the arrival of his

fellows. He was, as far as we know, the last of his species. The golden toad has not been seen since.

Other species at Monteverde were also affected. Two species of lizard vanished entirely. Today, the mountain's rainforests continue to be stripped of their jewels, with many reptiles, frogs and other fauna becoming rarer by the year. While still verdant enough to justify its name, the Monteverde Cloud Forest Preserve is beginning to resemble a crown that has lost its brightest gems.

Researchers began to study the records of temperature and rainfall. Eventually, in 1999, they announced that they had solved the mystery of the disappearance of the golden toad.

Ever since Earth passed through its first climatic magic gate in 1976, there were more and more mistless days each dry season on Monteverde, until they had joined into runs of mistless days. By the dry season of 1987, the number of consecutive mistless days had passed some critical threshold. Mist, you see, brings vital moisture. Its absence caused catastrophic changes.

Why, the researchers wanted to know, had the mist left Monteverde? Beginning in 1976 the bottom of the cloud mass had risen until it was above the level of the forest. The change had been driven by the abrupt rise in sea surface temperatures in the central western Pacific. A hot ocean had heated the air, elevating the condensation point for moisture. By 1987 the rising cloud-line was, on many days, above the mossy forest altogether, bringing shade but no mist. The golden toad has porous skin, and it likes to wander in daylight hours. It was exquisitely vulnerable to the new drier climate.

It's always devastating when you witness the extinction of a species. You are seeing the dismantling of ecosystems and irreparable genetic loss. It takes hundreds of thousands of years for such species to evolve.

The golden toad is the first documented victim of global warming. We killed it with our reckless use of coal-fired electricity and our huge cars, just as surely as if we had flattened its forest with bulldozers.

Since 1976 many researchers have observed amphibian species vanish before their eyes without being able to determine the cause. New studies indicate that climate change is responsible for these disappearances too.

In the late 1970s, a remarkable creature known as the gastric brooding frog disappeared from the mossy forests of southeastern Queensland. When it was first discovered in 1973 this brown, medium-sized frog astonished a researcher who looked into a female's open mouth – to observe a miniature frog sitting on her tongue! Not just the frog – scientists around the world were open-mouthed too.

The species is not a cannibal. It has bizarre breeding habits. The female swallows her fertilised eggs, and the tadpoles develop in her stomach until they metamorphose into frogs, which she then regurgitates into the world.

When this novel method of reproduction was announced, some medical researchers understandably got excited. How did the frog transform its stomach from an acid-filled digesting device into a nursery? The answer might help doctors treat a variety of stomach complaints. Alas, they were unable to carry out many experiments, for in 1979 – six years after humans learned of its existence – the gastric brooding frog vanished, and with it went another inhabitant of the same streams, the day frog. Neither has been seen since.

In the early 1990s, frogs began to disappear en masse from the rainforests of northern Queensland. Today some sixteen frog species (13 per cent of Australia's total amphibian fauna), have experienced falls in numbers. The decreases in rainfall experienced in eastern Australia over the past few decades cannot have been good for frogs. At least in the case of the gastric brooder and day frog, climate change is the most likely cause of their disappearance.

Now almost a third of the world's 6000-odd species of amphibians is threatened with extinction. Some scientists believe shallower breeding ponds – due to El Niño-like conditions – may be to blame. Fungal diseases are also contributing to the extinctions, and climate change is altering conditions in such a way that the fungus is flourishing.

Climate change seems to be the hidden cause of this wave of amphibian extinction.

The Three Tipping Points

Scientists are aware of three big tipping points for Earth's climate: a slowing or collapse of the Gulf Stream; the death of the Amazon rainforests; and the explosive release of methane from the sea floor.

All three occur in the virtual worlds of the computer models, and there is some geological evidence for all having happened at various times in Earth history. Given the current rate and direction of change, one, two or perhaps all three may take place this century. So what leads to these sudden shifts, what are the warning signs, and how might they affect us?

SCENARIO 1

Collapse of the Gulf Stream

The importance of the Gulf Stream to the Atlantic rim countries is enormous. In 2003 the Pentagon commissioned a report outlining the implications for US national security should the Gulf Stream collapse. The purpose of the report was, its authors said, 'to imagine the unthinkable'.

In their scenario the Gulf Stream slows as a result of fresh water from melting ice pouring into the North Atlantic. The planet continues to warm until 2010, but then a dramatic shift will occur – a 'magic gate' that will abruptly alter the world's climate.

The Pentagon 'weather report' for 2010 predicts persistent drought over critical agricultural regions, and a plunge in average temperatures of more than 3°C for Europe, just under 3°C for North America, and 2°C increases for Australia, South America and southern Africa.

The report predicts that nations won't co-operate with each other in the face of the disaster: mass starvation would be followed by mass emigration. Regions as diverse as Scandinavia, Bangladesh and the Caribbean would become incapable of supporting their populations. New political alliances would be forged in a scramble for resources. And war would be likely.

By 2010–20, with water supplies and energy reserves strained, Australia and the US would focus increasingly on border protection to keep out the migrating hordes from Asia and the Caribbean. The European Union, the report says, may go one of two ways – either it would focus on border protection (to keep out those homeless Scandinavians, among others), or be driven to collapse and chaos by internal squabbling.

In 2004 the Hollywood disaster movie *The Day after Tomorrow* also imagined the consequences of the shutdown of the Gulf Stream. For dramatic effect the time-lines for the collapse are greatly compressed, and the changes are far grander even than those imagined in the Pentagon report.

Scientists meanwhile have been working at understanding the consequences for biodiversity if the Gulf Stream were to collapse. They *are* catastrophic. If the currents will no longer carry oxygen into deeper waters, biological productivity in the North Atlantic will fall by 50 per cent, and oceanic productivity worldwide will decrease by over 20 per cent.

So what are the chances of the Gulf Stream shutting down this century? What would be the warning signs?

The Gulf Stream is the fastest ocean current in the world, and it is complex, spreading out into a series of spirals and sub-currents as its waters move northward. The volume of water in its flow is simply stupendous. You will recall that ocean currents are measured in Sverdrups, and one Sverdrup equals a flow of 1 million cubic metres of water per second. Overall, the flow rate of the Gulf Stream is around 100 Sverdrups, which is 100 times as great as that of the Amazon.

In its northern section the Gulf Stream is far

warmer than the waters that surround it. Between the Faeroe Islands in Denmark and Great Britain, for example, the Gulf Stream is a balmy 8°C, yet the surrounding waters are at zero. The source of the Gulf Stream's heat is the tropical sunlight falling on the mid-Atlantic, and the current is a highly efficient way to transport it, for one cubic metre of water can warm 3000 cubic metres of air.

In the North Atlantic, where the Gulf Stream releases its heat, it warms Europe's climate as much as if the continent's sunlight were increased by a third.

As the waters of the Gulf Stream yield their heat they sink, forming a great mid-oceanic waterfall. This waterfall is the powerhouse of the ocean currents of the entire planet, but history shows us that it has been interrupted many times.

Fresh water disrupts the Gulf Stream because it dilutes its saltiness, preventing it from sinking and thus disrupting the circulation of the oceans worldwide. Several Sverdrups or more of freshwater flow is needed. If the frozen north melted it could achieve that liquid potential, and to this we must add the increasing rainfall across the region.

The tropical Atlantic is becoming saltier at all depths, while the North and South polar Atlantic are becoming fresher. The change is due to increased evaporation near the equator and enhanced rainfall near the Poles. When similar changes were observed in other oceans, scientists realised that something – most probably climate change – had accelerated evaporation and rainfall rates over the oceans by 5 to 10 per cent.

This increasing tropical saltiness could lead to a temporary quickening of the Gulf Stream before its shutdown. Extra heat would be transferred to the Poles, which would melt more ice until enough fresh water could flow into the North Atlantic, collapsing the system altogether.

How fast might it happen? Ice-cores from Greenland indicate that, as the Gulf Stream slowed in the past, the island experienced a massive 10°C drop in temperature in as little as a decade. Presumably, similarly rapid changes were also felt over Europe, although no detailed record of climate has survived to tell of it.

It is possible, if the Gulf Stream were to slow, that extreme falls in temperature could be felt over Europe and North America within a couple of winters.

When is such an event likely to happen? Some climatologists think they are already seeing early signs of a shutdown. Not all agree – scientists at the Hadley Centre in England rate the chance of major disruption to the Gulf Stream this century at 5 per cent or less. Their main concern is an event in the Amazon that could be even more catastrophic.

SCENARIO 2

Collapse of the Amazon rainforests

One of the Hadley Centre's computer models is known as TRIFFID (Top-down Representation of Interactive Foliage and Flora Including Dynamics). It suggests that, as the concentration of CO_2 in the atmosphere increases, plants – particularly in the Amazon – start behaving in unusual ways.

The plants of the Amazon effectively create their own rainfall – the volume of water they transpire is so vast that it forms clouds whose moisture falls as rain, only to be transpired again and again.

But CO_2 does odd things to plant transpiration. Plants, of course, generally don't wish to lose their water vapour, as they have gone to some trouble to convey it from their roots to their leaves. Inevitably they do lose some whenever they open the breathing holes (stomata) in their leaves. They open their stomata to gain CO_2 from the atmosphere, and they will keep them open only as long as required.

Thus, as CO_2 levels increase, the plants of the Amazonian rainforest will keep their stomata open for briefer periods, and transpiration will be reduced. And with less transpiration there will be less rain.

TRIFFID indicates that, by around 2100,

levels of CO_2 will have increased to the point that Amazonian rainfall will reduce dramatically, with 20 per cent of that decline attributable to closed stomata. The rest of the decline, the model predicts, will be due to a persistent drought that will develop as our globe warms.

The current average rainfall in the Amazon of 5 millimetres per day will decline to 2 millimetres per day by 2100, while in the northeast it will fall to almost zero. These conditions, combined with an average rise in temperature of 5.5°C, will make the collapse of the Amazonian rainforest inevitable. A small change in temperature is capable of turning soils from absorbers of CO_2 to large-scale producers. As the soil warms, decomposition accelerates and lots of CO_2 is released. This is a classic example of a positive feedback, where increasing temperature leads directly to a vast increase in CO_2 in the atmosphere, which further increases temperature. With the loss of the rainforest canopy, soils would heat and decompose more rapidly, which would lead to the release of yet more CO_2.

This means a massive disruption of the carbon cycle. The storage of carbon in living vegetation would fall by 35 gigatonnes, and in soil by 150 gigatonnes. These are huge figures – totalling around 8 per cent of all carbon stored in the world's vegetation and soils!

The outcome of this series of positive feedback loops is that by 2100 the Earth's atmosphere would have close to 1000 parts per million of CO_2. Remember our current level is 380 parts per million and we need to act now to stop it reaching 550 parts per million.

This modelling experiment predicts devastation in the Amazon Basin. Temperatures rise by $10°C$. Most of the tree-cover is replaced by grasses, shrubs, or at best a savannah studded with the odd tree. Large areas become so hot and blighted that they cannot support even this reduced vegetation, and turn into barren desert.

When might all of this happen? If the model is correct we would start to see signs of Amazonian rainforest collapse around 2040.

By the end of the century the process would be complete. Half of the deforested region will turn to grass, the other half to desert.

What is so terrifying about this scenario is that climate change in the Amazon would itself hasten further runaway global climate change.

SCENARIO 3
Methane release from the sea floor

Clathrates is the Latin word for 'caged' and the name refers to the way ice crystals trap molecules of methane. Clathrates are also known as the 'ice that burns'. They contain lots of gas under high pressure, which is why pieces of the icy substance hiss, pop and, if lighted, burn when brought to the surface.

Massive volumes of clathrates lie buried in the seabed right round the world – perhaps twice as much in energy terms as all other fossil fuels combined. The clathrates in the seabed are kept solid only by the pressure of the cold overlying water. There are masses of clathrates in the

Arctic Ocean, where temperatures are sufficiently low, even near the surface, to keep them stable.

It's illustrative of the endless ingenuity of life that some marine worms survive by feeding on the methane in clathrates. They live in burrows within the icy matrix, which they mine for their energy requirements. There are between 10,000 and 42,000 trillion cubic metres of the stuff scattered around the ocean floor, compared with the 368 trillion cubic metres of recoverable natural gas in the world. It's not surprising that both worms and the fossil fuel industry can see a future in this weird material.

If pressure on the clathrates were ever relieved, or the temperature of the deep or Arctic oceans were to increase, colossal amounts of methane could be released. Palaeontologists are now beginning to suspect that the unleashing of the clathrates may have been responsible for the biggest extinction event of all time 245 million years ago.

At that time around nine out of ten species living on Earth became extinct. Known as the

Permo-Triassic extinction event, it destroyed early mammal-like creatures, thus opening the way for the dominance of the dinosaurs.

Many people think the cause of the extinction may have been a massive outpouring of lava, CO_2 and sulphur dioxide from the Siberian Trap volcanoes (the largest known flood basalt area). This would have led to an initial rise in global average temperature of about $6°C$ and widespread acid rain, which would have released yet more carbon. The increasing temperature then triggered the release of huge volumes of methane from the tundra and from clathrates on the sea floor. The explosive power to change climate would have been beyond imagination.

Two of these scenarios – the Amazonian dieback and the release of the clathrates – involve positive feedback loops, where changes build on each other to produce even greater changes. But there's one other positive feedback loop that's already occurring, and may be the trigger for further change.

Throughout our history we have engaged

in a constant battle to maintain a comfortable body temperature, which has been very costly in terms of time and energy. Just think of the hundreds of slight shifts in body position we make every day and night, and the taking off and putting on of overcoats and hats. Purchasing a house, our greatest personal expense, is primarily about regulating our local climate. In the US, 55 per cent of the total domestic energy budget is devoted to home heating and air conditioning. Home heating alone costs Americans US$44 billion per year.

As our world becomes more uncomfortable because of climate change, the demand for air conditioning will increase. In fact, during heatwaves it could mean the difference between life and death. But, unless we change how we create electricity, that demand for air conditioning will be met by burning more fossil fuels, which is a powerful positive feedback loop.

As global warming speeds up we will huddle at home clutching the remote of our climate control system, releasing ever more greenhouse gases. There is already a huge demand for air conditioners in countries such as the US and Australia where, until recently, construction codes for houses have been appallingly lax in regard to energy use.

Will we, in order to cool our homes, end up cooking our planet? Will air conditioning be one of the causes of the collapse of the Amazon or the interruption of the Gulf Stream?